Active Learning Astronomy™ for *Astronomy: The Evolving Universe*

9th Edition

Michael Zeilik

The University of
New Mexico

CAMBRIDGE
UNIVERSITY PRESS

PUBLISHED BY THE PRESS SYNDICATE OF THE UNIVERSITY
OF CAMBRIDGE
The Pitt Building Trumpington Street, Cambridge,
United Kingdom

CAMBRIDGE UNIVERSITY PRESS
The Edinburgh Building, Cambridge, CB2 2RU, UK
40 West 20th Street, New York, NY 10011-4211, USA
477 Williamstown Road, Port Melbourne, VIC
    3207, Australia
Ruiz de Alarcón 13, 28014 Madrid, Spain
Dock House, the Waterfront, Cape Town 8001,
    South Africa

http://www.cambridge.org

First published 2002

Printed in the United States of America

ISBN: 0-521-52901-8

Opinions expressed are those of the author and not
necessarily of the Foundation.

Cover design, book design, and type composition by John Cole
GRAPHIC DESIGNER, Santa Fe, NM

Front cover image
Magnificent Messier 82 (sometimes called the Cigar galaxy) cap-
tured by the Subaru Telescope on Mauna Kea, Hawaii, U.S.A. This
irregular galaxy lies about 12 million light years away in the constellation Ursae Major. Messier 82 is the brightest galaxy in the sky
in infrared light. Its core appears to have been disrupted by a close
encounter with another galaxy, triggering a colossal round of star-
birth. Winds from these hot, young stars expel supersonic hydro-
gen gas (red filaments) from the galaxy. Copyright 2000 National
Astronomical Observatory of Japan.

CONTENTS

Foreword for the Student

Focused Discussion Groups

Your course will promote cooperative learning by having Focused Discussion Groups during class. Research in cooperative learning over many years has shown that students at *all* levels learn science most effectively in small groups that have guided interactions. You will work at intervals in your groups, in which we'd like to have four or five people with three social roles: Reporter, Leader, and Skeptic. With groups greater than three, you will have more than one skeptic (generally a good idea!). You keep these roles during any one class, but you must switch them for other classes. That way, everyone will assume different responsibilities for the activities.

What do these social roles involve? Briefly,
- ❏ The REPORTER writes coherent reactions and summaries to be turned in.
- ❏ The LEADER works to start the discussion and keep it on track.
- ❏ The SKEPTIC tries to find holes in any arguments and reasoning.

Your instructor may elaborate on these roles or may even add other specific ones.

Each person in the group will write his/her name on the materials handed in by the Reporter, but each person should also write down the basic responses on his/her own sheet. The Reporter also has the responsibility of giving feedback to the group after the materials are returned by the instructor.

During class, your instructor will stop presenting and ask you to react (say to a demonstration or video), or answer a question, or participate in an activity from this book (or another source). Your group should then work together to accomplished the assigned task, within any time limit set by the instructor.

Task: To answer or react to the instructor's posed problem or activity.
Cooperative Learning: Create a group consensus response by
❑ Each person *formulating* his or her answer.
❑ Each person *sharing* his or her thoughts and responses.
❑ Everyone *listening carefully* to the comments of each other.
❑ The group creates a NEW answer/response that is better than the initial, individual formulations—by association, by building on each other's thoughts, and synthesizing a final result.
Criterion for Success: Each student must be able to explain the consensus answer or result.

Accountability: Your instructor will ask groups at random for their answers/responses. Any person in the group may be asked to give an oral explanation of the group's results. The materials handed in by the Reporter will become part of the evaluation by the instructor.

Note that you will need to exercise appropriate interpersonal skills for effective cooperative learning. These include: 1) trust, 2) clear communication, 3) acceptance and support of each other as unique individuals, and 4) constructive resolution of conflicts. You should reflect upon these skills after every group activity, no matter what your social role.

Table 1 lists some of the responsibilities expected of each social role.

Table 1. Sample Social Role Interactions for Cooperative Learning

Interactions	Examples
Leader	
Starts discussion.	*"Let's start with this idea. What do you think?"*
Directs sequence of steps.	
Keeps group "on track."	*"We'll come back later if we have time."*
Insures everyone participates.	
Checks the time.	*"We need to move on to the next step."*
Reporter	
Acts as a scribe for the group.	*"Do we understand this point?"*
Checks for understanding by all.	*"Explain to us what you think."*
Insures all members agree.	*"Do we all agree on this?"*
Summarizes conclusions.	*"So here's what we've decided…"*
Skeptic	
Helps group avoid coming to agreement too quickly.	*"What other possibilities are there?"*
Insures that all possibilities are explored.	*"Let's look at this a new way."*
Suggests alternative ideas.	*"I'm not sure we're on the right track."*
Looks for errors in reasoning.	*"Why?"*

6

Focused Discussion: Angular Size and Distance

Leader: _____   Reporter: _____
Skeptic: _____   Skeptic: _____   Skeptic: _____

Purpose: To investigate the relationship among angular size, actual size, and distance.
Evolving Universe **Connections:** p. 4, Fig. 1.2, Section 1.5, and Focus 1.2; *Concept Cluster*: Cosmic Distances.

Procedure: Attached is a photo of hot air balloons (Figure 1). From this image, and other information that you may have to find, answer the following questions:

1. Which of the balloons is closest? Label it C. The farthest? Label it F. *How do you know*?

2. How does the distance to balloon C compare to that of F? (Hint: Take a ratio!) What assumption do you have to make in order to reach your answer?

3. How can you find the *actual* distances between these balloons (say in meters)? (Hint: A typical hot air balloon has a diameter of about 20 meters when inflated.)

4. As a graph, sketch here the proper functional relationship between angular diameter and distance for the same object at different distances.

Concept Extension

What application does this concept have in astronomy, especially in the solar system?

Figure 1

Focused Discussion: Angular Speed and Distance

Leader: Reporter:

Skeptic: Skeptic: Skeptic:

Purpose: To be able to estimate the distance to an object from its angular speed.

***Evolving Universe* Connections:** p. 4, p. 9, p. 12, Table 1.2, p. 16, and Focus 1.2; *Concept Cluster*: Heavenly Motions.

Procedure: Examine the two photographs (Figures 1 and 2) of a jet airliner, which is so far away that the airplane is not visible, only its contrail. The photos were taken 20 seconds apart with a telephoto lens on a 35 mm camera. The images are enlarged so that $1° = 20$ mm on the prints.

1. What is the angular distance, in degrees, that the plane has traveled between the two photos?

2. What is the ratio of *distance traveled* to *distance from you* for the airplane? (Hint: An angle of 1° is a ratio of 1/57.)

3. What is the actual distance, in kilometers, that the plane has traveled between the two photos? (Hint: A commercial jet airliner flys at speeds of about 200 to 300 meters per second.)

4. Use #2 and #3 to estimate the airplane's distance. Does your result seem reasonable? What assumption(s) have you made to reach your result?

Concept Extension

How can you use this technique to find the distance to a nearby star?

Figure 1

Figure 2

Focused Discussion: Major Motions of the Planets

Leader:	Reporter:	
Skeptic:	Skeptic:	Skeptic:

Purpose: To describe the positions of the sun and selected, naked-eye planets along the ecliptic and to infer general patterns in their motions from graphs.

***Evolving Universe* Connections:** Section 1.4, Table 1.2, Focus 1.2; *Concept Clusters*: Cosmic Distances and Heavenly Motions.

Background skill: Reading and finding slopes of graphs.

Procedure: You have two types of graphs (Figures 1 and 2) showing the positions of the planets. Both indicate time running down the page; the dates at the left are given in 10-day intervals (the start of a year is given on the right). The vertical extent covers a little more than a year, and the horizontal span is 360°. The line with the solid dot represents the motion of the sun, which is labeled first.

One set of graphs is labeled "Planet Positions Relative to the Sun" (Figure 1). Here the sun's position is fixed and runs down the middle of the page. Note that the planets are visible in the evening sky to the *left* of the sun's line, and in the morning sky to the *right*. When seen in the evening sky, a planet is *east* of the sun; it sets *after* the sun sets. When seen in the morning sky, a planet is *west* of the sun; and it rises *before* the sun rises. When a planet crosses the sun's line from left to right, it moves in visibility from the evening sky to the morning sky.

Note that the angular distances of the planets from the sun change with time. The angular distance between the sun and a planet on any given date is called the planet's *elongation*. If you scan down the graph, you should notice that certain planets stay near the sun; others don't.

In the "Zodiacal Graph" (Figure 2), the planetary positions run across the page horizontally to the left through the constellations of the *zodiac* (see the diagram at the top of the page; each constellation of the zodiac is drawn and labeled; can you identify any?). Once around the zodiac is a complete circle of 360°. The solid line through the center of the constellations is the path of the sun in the sky relative to the stars—the *ecliptic*. Note that the constellations of the zodiac are those that lie along the ecliptic.

The legend for each line indicates the planets plotted and the year. Note that some planets take longer than others to complete one circuit of the zodiac. Those that do so quickly have a faster *angular speed* than the others. That is, they move a certain angle on the sky in less time, or for the same time interval, they move a larger angle.

Part I. Examine Figure 1 first. You need to determine the scale, so that you know that 1 mm = x degrees on the paper. What is your value for x? (Hint: How many degrees span the graphs horizontally?)

_____

1. Which planet always appears close to the sun in angular distance?

2. Over the course of a year, how often does Mercury move from the morning to the evening sky? From the evening to the morning sky?

3. Is this change more or less frequent for Mars? On about what date does Mars move from the morning to the evening sky?

4. For Mercury, does its maximum angular distance from the sun remain constant or does it vary? (This maximum distance is called *maximum elongation*.) Determine angular distance for at least three dates and write down the average *in degrees*.

Part II. Now examine Figure 2. It looks very much like the previous one, except now the positions, including that of the sun, are plotted relative to the stars of the zodiac. Note that *east* is to the *left*, and *west* is to the *right*.

5. How does the sun's motion differ in this graph compared to those of the planets? In what direction does the sun move relative to the stars? How long does it take to move 360°? What have you calculated (look at the units)? How does this quantity relate to the *slope* of the sun's line?

6. How does the sun's motion resemble *in general* that of the planets?

Figure 2. Zodiacal Graph

Figure 1. Planet Positions Relative to the Sun

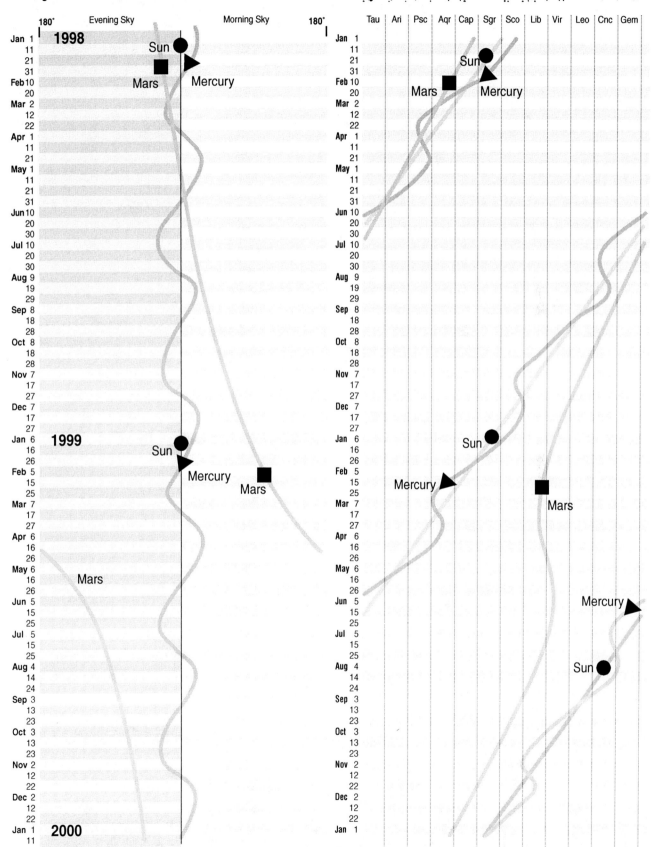

© 2002 Michael Zeilik

13

7. How does the sun's motion *differ from* that of the planets?

8. Note the cases when the planet's motion is toward the *right (west)* rather than to the east. During these times, a planet's motion is called *retrograde*. What is the duration, in days, of the retrograde motion of Mars? Of Mercury?

9. What is the average angular speed of Mercury eastward?

Concept Extension

Consider the planets Venus and Jupiter. Which one would have motions most like Mercury? Most like Mars? Why?

Focused Discussion: Retrograde Motion of Mars

Leader:	Reporter:	
Skeptic:	Skeptic:	Skeptic:

Purpose: To explore the characteristics of one retrograde motion of the planet Mars.
***Evolving Universe* Connections**: pp45–46, Figure 3.4; Concept Cluster: Heavenly Motion.

Figure 1. 1992-1993 Retrograde Motion of Mars

Gemini — West — Mars (10 day interval for large squares) — East

Procedure: If you were to plot the position of a planet against the background of the stars night after night, you would notice that the planet usually moves from west to east (*eastward*) across the sky over the course of several weeks. From time to time, however, a planet will appear to slow in its eastward progression, reverse its motion and move for a period of time *westward*, then again slow and reverse directions, resuming its normal eastward path. This apparent reversal in the planet's course is called *retrograde motion*, the part of the motion that is westward.

Figure 1 represents this motion for the planet Mars, as seen from earth, relative to the stars. TRACE OVER THE RETROGRADE PART OF THE LOOP.

Now let's look at an overview of retrograde motion for any two moving bodies, one moving faster than the other. Figure 2 provides you with two orbits, each with points corresponding to the positions of the two orbiting bodies in equal intervals of time. Note that the middle point (6) has been started for you. Extend the arrow to the right and number it. Connect every other point (8, 10, 12; then 4, 2, 0) of the inner orbit to its corresponding point on the outer orbit with a *straight* line, extending the line *all the way to the right of the page*. Number each line for each pair of points at the far right. These numbers show the position of the outer planet, as seen by the inner, against the background of the stars. Note that the inner planet moves faster than the outer one.

Figure 2. Retrograde Motion of Two Planets

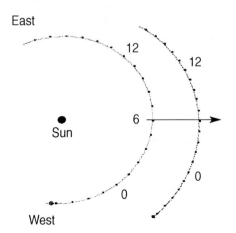

East

12   12

6 →

Sun

0   0

West

1. Using the diagram you just completed, explain why the outer planet appears to move retrograde as seen from the inner planet. (Hint: What happens to the line of sight relative to the stars?)

2. At what position Figure 2 is the outer planet in the *center* of its retrograde motion? What is its orientation relative to the sun as seen from the earth?

3. Now imagine you are standing on the outer planet and observing the inner one. What would you see? (Hint: Think about #2 for the inner planet.)

Concept Extension

Consider Jupiter and the earth. How would an *explanation* of Jupiter's retrograde motion differ from that of Mars'? (Hint: How would you revise Figure 2?)

Focused Discussion: Sunrise Points

Leader:	Reporter:	
Skeptic:	Skeptic:	Skeptic:

Purpose: To discover the seasonal variation of sunrise points.

***Evolving Universe* Connections:** pp. 8-9, Figure 1.8; *Concept Cluster*: Heavenly Motions.

Procedure: Your answer to that question would most likely be "in the east, right?" Well, two days of the year, you would be exactly correct. The rest of the year, however, things are a bit more complicated.

Table 1 gives the sun's position in degrees of azimuth, which is the angle from north, that is, due north is 0°, due east is 90°, due south is 180°, and due west is 270°. Plot each sunrise on Figure 1, labeling each position with its date (early spring = esp, midsummer = msu, and so on).

Table 1. Sunrise Points Table

	Sunrise Angle (degrees)		Sunrise Angle (degrees)
Early Spring	91.4	Early Autumn	91.6
Mid Spring	84.8	Mid Autumn	101.3
Late Spring	74.8	Late Autumn	113.7
Early Summer	61.2	Early Winter	120.7
Mid Summer	72.5	Mid Winter	110.5
Late Summer	83.9	Late Winter	101.7

Figure 1. Horizon Profile

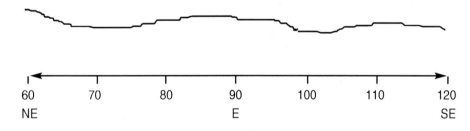

1. On what two days does the sun rise closest to due east? At what times of the year does the sun rise *due east?*

2. When does the sun rise farthest south, and when does it rise farthest north? At what times of the year does this occur?

3. If the sun rises at its farthest south point one morning, will the day be long or short? Why?

4. How would the sunset points differ from or be similar to the sunrise points? (Hint: Imagine Figure 1 gave a profile of the western horizon, centered on 270°.)

Concept Extension

If you were observing the sun from the earth's *southern* hemisphere, how would the data in the table differ?

Focused Discussion: Solar System Models

Leader: _____ Reporter: _____

Skeptic: _____ Skeptic: _____ Skeptic: _____

Purpose: To relate naked-eye observations to a geocentric and heliocentric model of the solar system.
***Evolving Universe* Connections**: Section 2.4, Table 2.1, Figure 2.11, Section 3.2; *Cosmic Clusters*: Cosmic Distances, Heavenly Motions, and Scientific Models.

Procedure: Attached are two blank templates for a "god's-eye" view looking down from above of the paths of selected planets in the solar system. NOTE THAT EASTWARD IS COUNTERCLOCKWISE. In the Geocentric Model (Template 1) mark the earth at the center with a large dot and label "Earth." In the Heliocentric Model (Template 2) mark the sun at the center with a large dot and label "Sun." You will need to position on each model the sun and planets viewed from the earth as follows:

It is just after sunset. Mars is rising in the east. Venus is at maximum eastern elongation, 45 degrees from the sun. Jupiter is halfway between the sun and Mars in angular distance.

1. What are the angular positions of the planets, relative to the sun, along the ecliptic? Consider the sun at 0 degrees, with angles increasing *counterclockwise.*

2. Start with the Geocentric Model. Which path belongs to which planet? Place the earth on the correct path, right on the horizontal line that cuts through the center of the template. Then place Mars in its proper position, then Jupiter, and then Venus. Note that Venus must have a special position along its path. Label all the planets.

3. Now turn to the Helio-centric Model. Which path belongs to which planet? Start with the sun; place it on its path right on the horizontal line that cuts through the center of the template. Then place Mars in its proper position, then Jupiter, and then Venus.

4. Compare the figures. Which model gives a *simpler* layout of the observations? (Hint: What do you think is meant by "simpler" in this context?)

Concept Extension

If you were to stand in the center of each model, would the angular positions of the planets differ? If so, how? If not, why not?

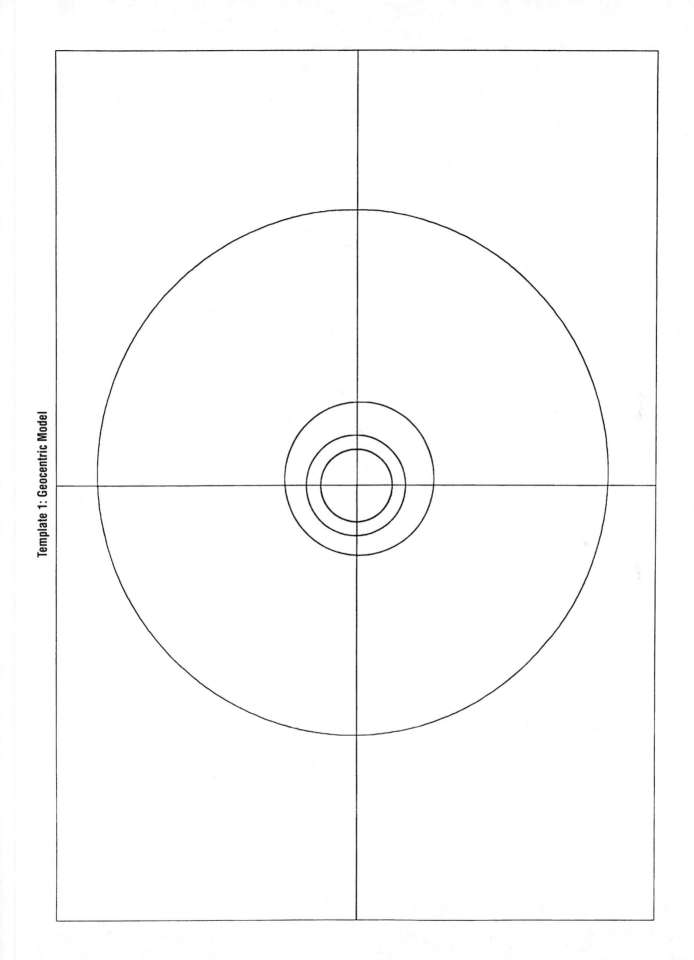

Template 1: Geocentric Model

21

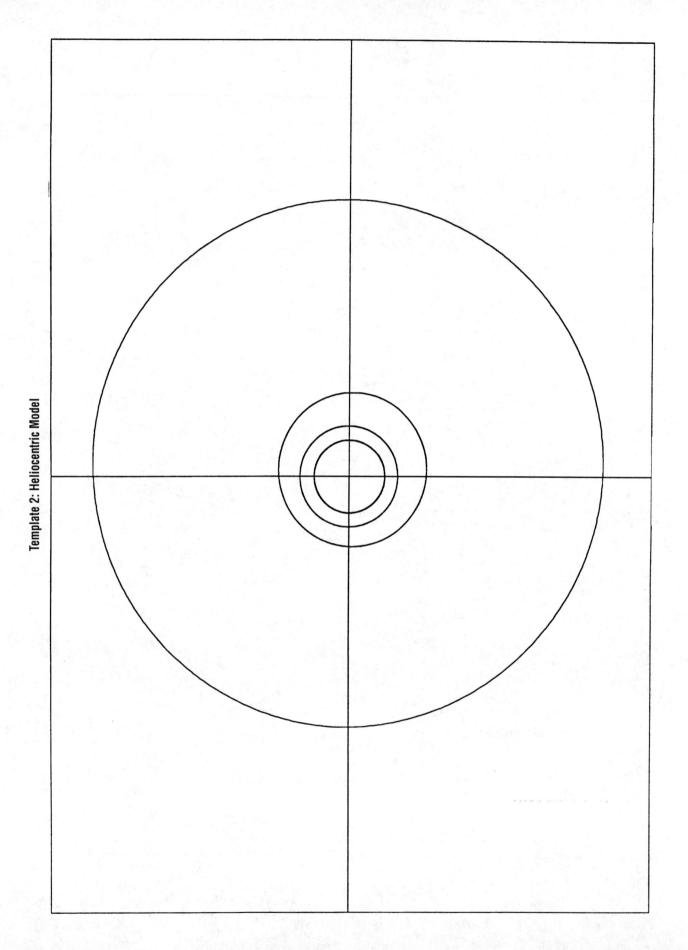

Template 2: Heliocentric Model

22

© 2002 Michael Zeilik

Focused Discussion: Kepler's Third Law—Planets

Leader: Reporter:

Skeptic: Skeptic: Skeptic:

Purpose: To use the motions of the planets to infer a fundamental pattern.

***Evolving Universe* Connections**: Section 3.5, Table 3.2, Figure 3.13; *Concept Clusters*: Cosmic Distances, Heavenly Motions.

Procedure: You have a table (see Table 1) of the five naked-eye planets and their average distances (in AUs) and orbital periods (in years) for their motions around the sun. Note the column in which the periods (P) are squared (#3) and the distances (a; semimajor axis of the elliptical orbit) is cubed (#5). In the next column (#6), you see the period squared divided by the distance cubed. The last column (#7) gives the average orbital speeds. This basic information comes from a heliocentric model of the solar system.
For each planet, compare the values in column #6.

---

1. Are they very similar or
very different? How so?

Now use Graph Template 1 to plot a point for each of the planets, using the distance (column #4, x-axis) against period (column #2, y-axis). Start with the earth—it's easy! Note that Graph Template 1 starts at 0.1 and goes 0.2, 0.3, *etc.*, until it hits 1.0. Then it goes from 1.0 to 10, and 10 to 100. Label each point by the planet's name. Can you draw a straight line through your plotted points? If so, do it!

---

2. Now imagine a body
found orbiting the sun at 3
AU. Where would it fall on
your graph? What, roughly,
would be its orbital period?

Uranus, Neptune, and Pluto were discovered with telescopes. Their distances are about 19, 30, and 40 AU. Add Pluto to your table (fill in the blank cells in Table 1) and to your graph.

---

3. Now imagine a new body
were found beyond Pluto at a
distance of 60 AU. What did
you predict about the value
of P^2/a^3 for this object?
Where would it fall on your
graph? What would be its
orbital period?

4. Look at Graph Template 2, in which are plotted the orbital speeds versus distance (column #7). Draw a smooth curve through all the points. How do the orbital speeds of the planets vary with distance from the sun? What would the orbital speed be of a body at 3 AU? Clearly mark its position on the graph.

Table 1. Orbital Properties of the Planets

1 Planet	2 Period (P; years)	3 P^2	4 Distance (a; astronomical units)	5 a^3	6 P^2/a^3	7 Average Orbital Speed (km/s)
Mercury	0.24	0.058	0.39	0.059	0.97	48
Venus	0.62	0.38	0.72	0.37	1.0	35
Earth	1.0	1.0	1.0	1.0	1.0	30
Mars	1.9	3.6	1.5	3.4	1.1	24
Jupiter	12	140	5.2	140	1.0	13
Saturn	29	840	9.5	860	0.98	10
Pluto			40			5

Concept Extension

Long-period comets have semi-major axes of some 50,000 AUs. What is their typical orbital period?

Graph Template 1

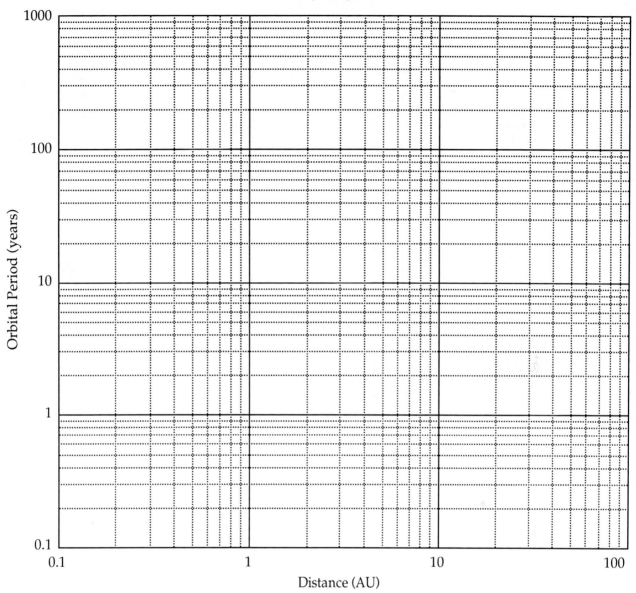

Orbital Period (years)

Distance (AU)

Graph Template 2

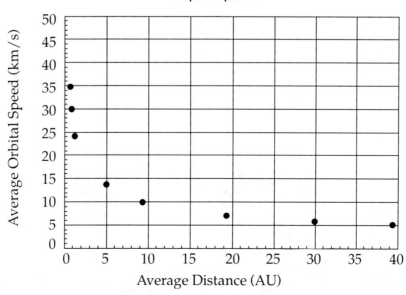

Average Orbital Speed (km/s)

Average Distance (AU)

Reflection

How well has your cooperative learning group been functioning? Consider and reach a consensus on each of the following:

1) Acceptance and trust
2) Clear communication
3) Support of individuals and ideas
4) Constructive criticism and resolution of conflicts

Has one person been dominating the thinking of the group? If so, how can this be avoided in the future?

Focused Discussion: Kepler's Third Law—Mass of Jupiter

Leader: _____   Reporter: _____

Skeptic: _____   Skeptic: _____   Skeptic: _____

Purpose: To use the motions of the moons of Jupiter to find Jupiter's mass.

Evolving Universe Connections: Section 3.5, Table 3.2, Figure 3.13; _Concept Clusters_: Cosmic Distances, heavely Motions.

Procedure: The motions of Jupiter's four largest (and brightest) moons are shown in Figure 1 in graphical form. Note the layout is horizontal. Three weeks are plotted for the paths of Jupiter's Galilean moons, labeled Io, Europa, Callisto, and Ganymede (in order outward from Jupiter). They are shown in proper relative distances from Jupiter, and the disk of Jupiter across the center is to the _same_ scale. The graph resembles a motion picture with all the frames drawn sequentially. For each vertical line, the dates are given every four days, from left to right.

How to find the mass of Jupiter from these motions? Apply Newton's version of Kepler's third law! You have seen it stated as P^2/a^3 = constant, where P is the orbital period and a the semimajor axis of the orbit.

But what's in that constant? Newton found, from his law of gravitation, that the constant contained the _sum_ of the orbiting mass and the mass about which the other body is orbiting. So for the earth and the sun, the constant contains $M_{sun} + M_{earth}$. Since the sun's mass is so much greater than that of any body orbiting it, we can essentially ignore the mass of the orbiting body, then $M_{sun} + M_{earth} \approx M_{sun}$. Hence, we can use Kepler's third law for any body orbiting the sun to find out the sun's mass! (_Warning_: if the two masses are comparable, such as in a binary star system, we can't use this approximation.)

The mass of any of the Galilean moons is much smaller than that of Jupiter, so we can use the orbital motions of any one of the moons with Kepler's third law. Let's use Ganymede. You need to find the orbital period (in days) and the distance (in Jupiter radii). For the period: find a point when Ganymede appears farthest away from the planet. Then look along the figure until you see that position again on the _same side of the planet_. The interval, in days, is the orbital period—the time for one complete revolution. You should measure to the nearest half day.

---

1. What is the orbital period of Ganymede in days? _____ days

---

2. For the distance: use a ruler to measure the diameter of Jupiter in _millimeters_ (the width of the horizontal line across the center of the figure). Do this at least three times and average the results!

What is Jupiter's diameter? _____ mm

3. Then measure the distance, in millimeters, *from the center of Jupiter*, to the farthest point of Ganymede orbit. (Use the center of Jupiter's line and the center of Ganymede's line as reference positions.)

What is Ganymede's maximum orbital distance? _____ mm

4. Now we'll do a series of calculations using this information. First, divide the value for Ganymede's orbital distance (in mm) by the value of Jupiter's diameter (in mm) to obtain the orbital size in "Jupiter diameters."

Jupiter diameters = radius of orbit = a = _____

Now cube this number: a^3 = _____

Take Ganymede's orbit period, P, in days, and square it. P^2 = _____

5. Now divide the period squared by the distance cubed. P^2/a^3 = _____

6. This value is the constant for any body orbiting Jupiter (but in funny units; note it is *not* 1)! Now to get the mass. We'll spare you all the details of converting from the units you used above to SI units. Divide the number 2.5 by the value you calculated in step 5. That will give you the mass of Jupiter in units of 10^{25} kg.

Mass of Jupiter = $2.5/(P^2/a^3)$ = _____ $\times 10^{25}$ kg (Two significant figures!)

Concept Extension
How can you find out the
mass of Saturn?

Figure 1. Motions of the Galilean Moons of Jupiter

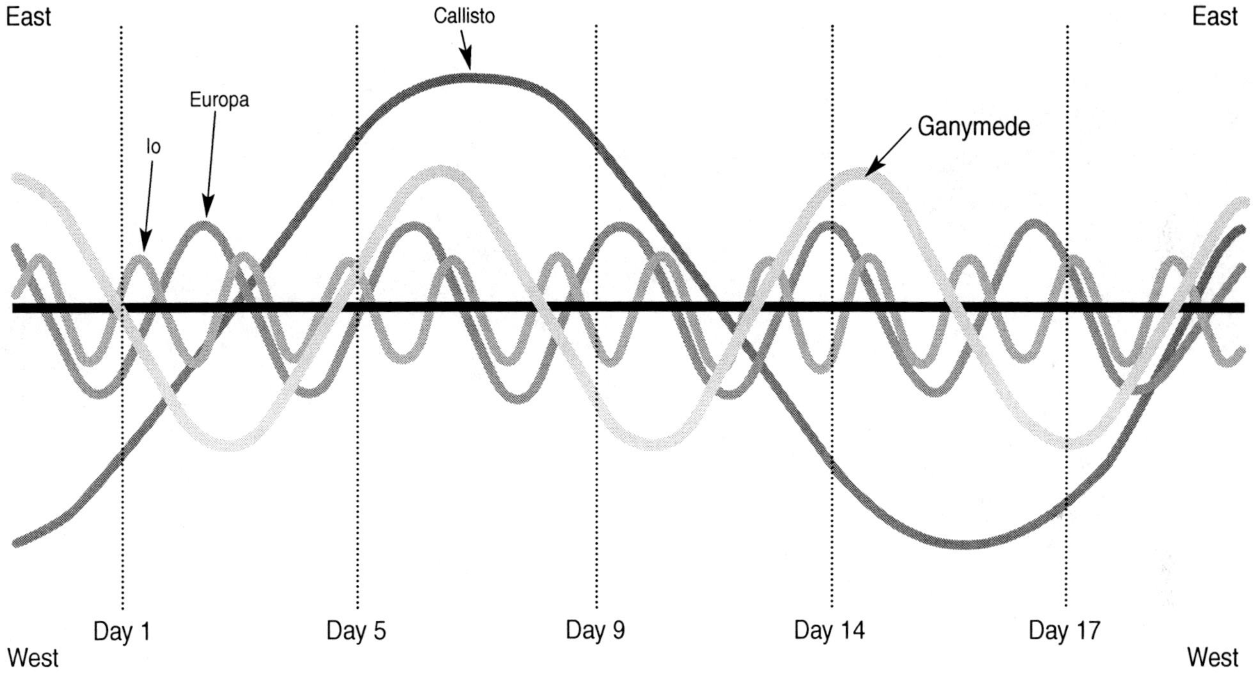

East

East

Callisto

Europa

Ganymede

Io

Day 1

Day 5

Day 9

Day 14

Day 17

West

West

Reflection

How well has your cooperative learning group been functioning? Consider and reach a consensus on each of the following:

1) Acceptance and trust
2) Clear communication
3) Support of individuals and ideas
4) Constructive criticism and resolution of conflicts

Has one person been dominating the thinking of the group? If so, how can this be avoided in the future?

Focused Discussion: Pluto and Charon—A Double Planet System

Leader:	Reporter:	
Skeptic:	Skeptic:	Skeptic:

Purpose: To find the individual masses of Pluto and Charon.

Evolving Universe **Connections**: Section 4.5, Focus 4.3, Figures 4.18, 4.19, and 10.29; *Concept Cluster*: Heavenly Motions

Procedure: Pluto and Charon make up a gravitationally-bound system whose motions follow Kepler's laws. From the recent series of eclipses, we know the orbital period very well: 6.387 days. Earlier observations showed that Pluto's brightness varied with a period that is the same as the orbital period. It is assumed that Pluto and Charon have tidally-locked rotation periods that are equal to their orbital periods.

1. Which of Kepler's laws must you use to find the masses?

2. The semimajor axis of the orbit is about 19,100 km. Use this information, the information given above, and the application of one of Kepler's laws to do the appropriate calculation. Watch out for your units!

3. You should be stumped here because you need an essential piece of information. What is it? Right, you need the position of the *center of mass*! Charon is five times Pluto's distance from the center of mass. Mark the approximate position of the center of mass on Figure 1, which is a scale model of the system. Now finish the calculation.

Concept Extension

How would you find out the individual masses of stars in a binary system?

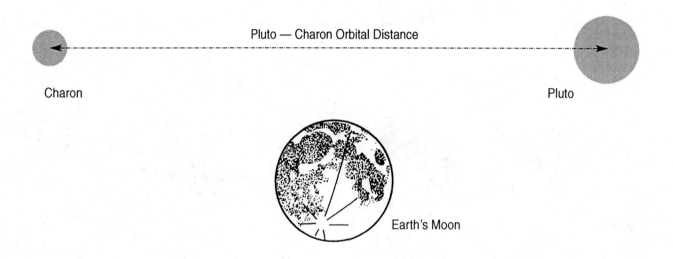

Figure 1. Pluto — Charon System to Scale

Focused Discussion: The Doppler Shift and Planets

Leader: _____ Reporter: _____

Skeptic: _____ Skeptic: _____ Skeptic: _____

Purpose: To apply the Doppler shift to the rotation of planets.

***Evolving Universe* Connections**: Focus 10.1; *Concept Cluster*: Celestial Light and Spectra.

Procedure: Imagine you are looking straight onto the equator of a rotating planet. Light emitted from (or radar waves bounced off of) the edge approaching you will be blue shifted; from the opposite edge, red shifted. If the planet has a rotational period P and a radius R, then the Doppler shift from either edge is related to these quantities by

$$\Delta\lambda/\lambda_0 = \Delta v/v_0 = 2\pi R/cP$$

where c is the speed of light.

Figure 1 shows the received signal of a radar pulse sent to Venus from an earth-based telescope. The signal was sent out at a frequency of 430 MHz (1 MHz = 10^6 Hz) and covered the entire angular diameter of the planet as seen from the earth. Note the signal has been spread out in frequency, relative to the center frequency (430 MHz). The peak at 0 Hz is the return signal from the center of the planet's disk; it is *not* Doppler shifted (frequency shift is zero). The peaks from the edges of Venus show a maximum blue shift and a red shift.

1. What is the maximum
blue shift, in hertz?

2. What is the maximum
red shift, in hertz?

3. Average these two values
to get a Doppler shift across
the disk of Venus.

4. Calculate the rotation period of Venus in days. For simplicity, use 6000 km for the radius of Venus and 3×10^5 km/s as the speed of light. Compare your value to the value in your textbook. How do they compare?

Concept Extension

How could you use this procedure to find the rotation period of Mercury?

Figure 1. Radar Pulse Reflected From the Surface of Venus

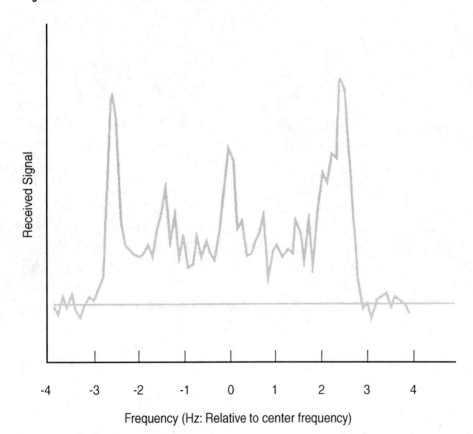

Focused Discussion: Weight

Leader: _____ Reporter: _____

Skeptic: _____ Skeptic: _____ Skeptic: _____

Purpose: To find out how the weight of a mass varies with its distance from the center of the earth.

***Evolving Universe* Connections:** pp. 78-80; *Concept Cluster*: Heavenly Motions.

Procedure: You have a table (see Table 1) of the weight of an object at various distances from the center of the earth. Note that the distances are measured in earth radii. The object weighs 100 kg at the surface. Using the Graph Template, plot these values, starting at 1 earth radius. After you have plotted all the points, try by hand to draw a smooth curve through all of them. When you have completed the graph, answer the following questions:

1. Consider the object located at 2 earth radii. What would be its weight?

2. What is the overall shape of the curve? What kind of relationship does it represent? (Hint: Direct, inverse or inverse square?)

3. Imagine you had the same mass at 10 earth radii. What would be its weight? (Hint: Use the relationship you inferred in #2.)

4. At a very, very great distance from the earth (or any mass), what would be the weight of the object?

Concept Extension

What is the difference between mass and weight?

Table 1. Weight versus Distance

Distance (earth radii)	Weight (kilograms)
1.0	100.0
1.5	44.0
2.0	25.0
2.5	16.0
3.0	11.0
3.5	8.2
4.0	6.3
5.0	4.0

Graph Template

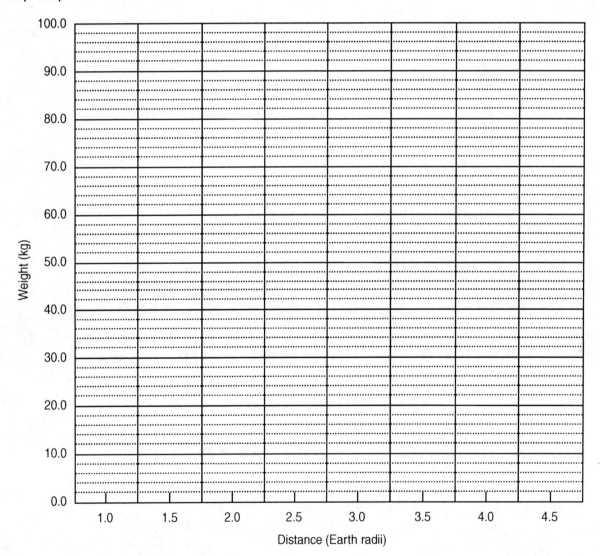

Focused Discussion: Newton's Gravitation

Leader: _____   Reporter: _____

Skeptic: _____   Skeptic: _____   Skeptic: _____

Purpose: To investigate Newton's concept of gravitation.

***Evolving Universe* Connections**: Section 4.4; *Concept Cluster*: Heavenly Motions.

Procedure: Below are four mental puzzles about gravitation. To solve them, you will need to extend concepts you already know. You can choose to support your argument mathematically, but if you do so, you must use words/pictures first.

1. Suppose the sun suddenly became very small but its mass remained the same. What would happen to the earth's orbit?

2. Imagine the earth in the form of a spherical *shell*, so that this shell contained the total mass of the earth. The size of the earth remains the same. What would be its gravitational force on a test mass placed (a) on the surface of the earth, and (b) inside the earth?

3. Imagine a completely smooth and spherical earth with no atmosphere. You have a baseball to place into an orbit just above the surface. What concept(s) would you apply to determine the period of the baseball's orbit? How would the period change if you used a bowling ball?

4. Imagine a straight shaft bored from the earth's surface, through the center of the earth, and out the other side. Drop a baseball down this shaft. Predict the baseball's motion (a) at the start, (b) at the center of the earth, and (c) just as it comes out the other side. At each point, describe the *acceleration* and *velocity* of the baseball. Will this motion be periodic? If so, how does the period compare to that in #3?

Concept Extension

Suppose you replaced the earth with Mars. How would your answers change in #1 to #4, if at all?

Focused Discussion: Continuous Spectra

Leader: Reporter:

Skeptic: Skeptic: Skeptic:

Purpose: To be able to identify, graph, and physically interpret continuous spectra.

Evolving Universe **Connections**: Section 5.2, Fiugre 5.11, pp. 257-259, Focus 12.2; *Concept Cluster*: Celestial Light and Spectra.

Procedure: A *spectrum* consists of visible light spread out over wavelength (or color, if you consider the visible spectrum only). When you look at a photo of a visible spectrum, your eye senses the colors that are emitted by the source. In an emission line spectrum, for example, you see bright lines at certain colors only, because these lines are the only wavelengths at which the source is emitting energy. Between the lines, where the source is not emitting energy, the spectrum appears dark. In contrast, a continuous spectrum shows a smooth band of colors as you view it. Your eye "measures" the wavelength by revealing the colors, from violet (shorter wavelengths) to red (longer wavelengths). Now, you are also "measuring" the energy emitted each second at each color—the *intensity*—but you have no easy way to quantify this property.

Imagine that you have a light meter that is equally sensitive to all colors. You place color filters between the source and the light meter and record the reading of the intensity. Table 1 shows a series of such readings from 3000 (UV) to 10,000 Ångstroms (infrared).

The values are all relative to the peak, so they indicate *relative intensity*. Note that the relative intensities vary over the range of wavelengths; this indicates to you that the source is *not* emitting the same energy per second at each wavelength.

Let's reveal this energy dimension by plotting the spectrum in the Graph Template. Plot the relative intensity at each wavelength given in Table 1. When done, try to draw as *smooth* a curve as possible through all the points.

Table 1. Relative Intensity Values for a Continuous Spectrum

Wavelength (Å)	Relative Intensity
3000	0.46
3500	0.70
4000	0.89
4500	0.96
5000	1.00
5500	0.98
6000	0.93
6500	0.85
7000	0.78
7500	0.70
8000	0.63
8500	0.56
9000	0.50
9500	0.45
10,000	0.40

Once you've completed your graph, answer the following questions:

1. Discribe the overall shape of your spectrum?

2. At what wavelength (in Ångstroms) does the spectrum have a peak?

If you compare the spectrum with those in your textbook, you will see that it has the characteristic shape of a *Planck curve*. This shape tells you that the emitting source is a hot *blackbody*.

___

3. What *one* physical property does the emission of a blackbody depend upon?

___

4. *Wien's law* provides the relationship between the wavelength at which a Planck curve peaks and the temperature of the blackbody. It is

Temperature = $2.9 \times 10^7 \div$ Peak Wavelength

where the temperature is in kelvins if the peak wavelength is in Ångstroms. Using your result from #2, calculate the temperature in kelvins of the blackbody whose spectrum you have drawn.

Calculated Temperature = _____

___

Concept Extension

Suppose you are measuring the spectrum of a blackbody with a temperature of 10,000 K.

5. Do you expect its peak to be at longer or shorter wavelengths than for the spectrum given here?

Graph Template

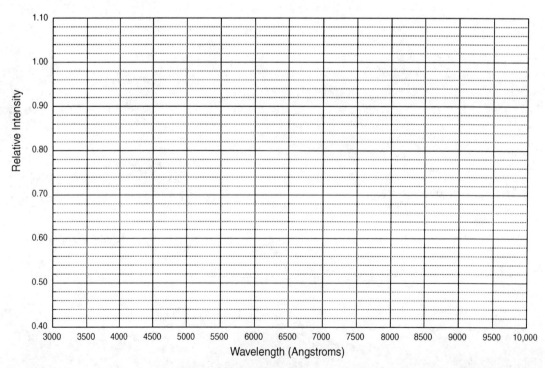

© 2002 Michael Zeilik

Focused Discussion: Stellar Temperatures, Colors, and Spectra

Leader: Reporter:

Skeptic: Skeptic: Skeptic:

Purpose: To examine stellar spectra and find out how temperature distinguishes them.

***Evolving Universe* Connections:** Sections 13.3 and 13.4; *Concept Cluster*: Celestial Light and Spectra.

Procedure: You will be given spectra of six stars (labeled A through F). Your task: to order them by temperature, from the hottest to the coolest. First, examine the spectra as a group.

1. In general, what kind of spectra do these stars display? Use Kirchhoff's rules to describe the physical conditions under which these spectra form.

2. Draw a freehand line though the spectra, smoothing out the wiggles in a way that traces out an a verage value. Which star is the hottest? The coolest? How do you know?

3. For each star, estimate its surface temperature, assuming that it radiates like a blackbody. Write your estimate in kelvins in Table 1.

Table 1

Star ID	Estimated temperature (kelvins)
A	
B	
C	
D	
E	
F	

4. Now sort the spectra in order of *decreasing* temperature. Write your sequence, by star ID, here. Which star would have a color most like the sun's?

5. **Optional Extension.**
Find two Balmer lines in the spectra. Note that the intensities (depths in these graphs) of these lines vary. What pattern do you see in the Balmer lines as you go from hotter to cooler stars? What color stars have weak or no Balmer lines?

6. **Optional Extension.**
Without identifying other lines, do you see any patterns in them? That is, do you find the same lines at the same wavelength positions? (The depth of the lines may vary.)

Concept Extension
You are given the spectrum of an unknown star. How could you estimate its surface temperature?

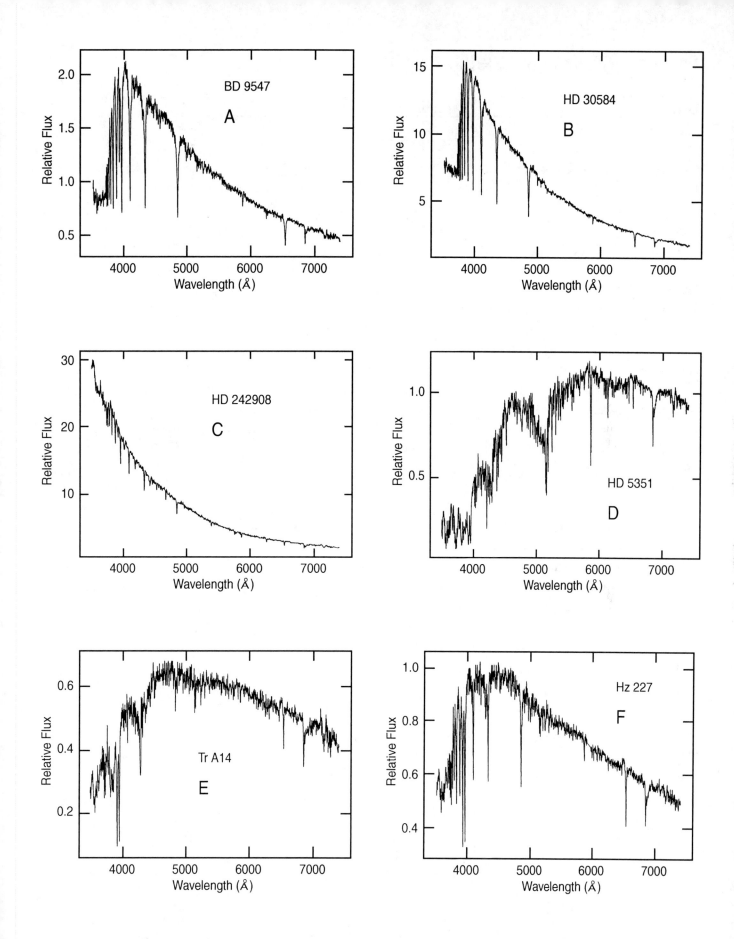

Adapted from *A Display Atlas of Stellar Spectra* (Dept. of Astronomy, University of Washington). Used with permission.

Focused Discussion: Stellar Parallax

Leader: _____ Reporter: _____

Skeptic: _____ Skeptic: _____ Skeptic: _____

Purpose: To gain a concrete understanding of parallax.

***Evolving Universe* Connections:** Section 13.2 and Focus 13.2; *Concept Cluster:* Cosmic Distances.

Procedure: What is parallax, and is it important only in astronomy? Not at all! Parallax is something we experience and use every day of our lives. We use it automatically, entirely unconsciously, many times each day. It is parallax that allows us to gauge depth, that important third dimension. We use our own built-in parallax detection system to judge how far away that cup of coffee is when we reach for it, whether or not we can cross the street safely, where the doorway is and the wall isn't! What is this system? Your own two eyes!

In stellar parallax, an apparent shift in the position of a star is observed when the earth moves from one side of the sun to the other (Figure 1). First, we observe the star from point *A* (say in January) and it appears to be at point *a* relative to background stars. Six months later (in June), we observe the star from point *B*, and it appears to be at point *b*. This is the key observation that demolishes the geocentric model. That the earth revolves around the sun allows us to observe this star from two different positions, so that it appears to shift from one position to another relative to background stars that are farther away.

Our own equipment makes observing easier. Since we have two eyes, we don't have to wait six months. We already have two different lines of sight to an object, one for each eye. Depth perception is the automatic mental processing of these two different lines of sight to produce an intersection. Our brain then tells us that it is at this point of intersection that the object is actually located. In analogy to Figure 1, if the sun is your nose, and the two positions of the earth six months apart (*A* and *B*) are your eyes, then the lines from **A** to **a** and **B** to **b** are the lines of sight from your eyes through the object. Your clever brain, which understands parallax very well indeed, then interprets this data to tell you that the object is where the star actually is.

Close one eye. Now try to reach out and touch the tip of a pen or pencil held in your other hand. It's a lot harder with one eye, isn't it?

Now you will explore the apparent shift that occurs when you are observing location shifts. Again consider your eyes to be the positions of earth six months apart. You will hold a pencil upright at several different distances, but always directly in front of your nose. Every group member will go through the procedure, so that each of you experiences the parallax shift for yourself.

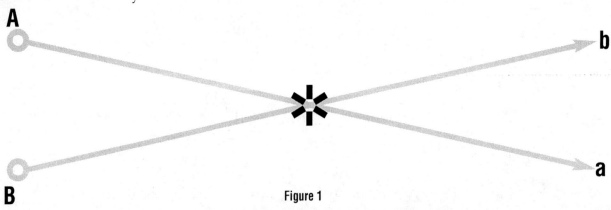

Figure 1

1. Hold the pencil right at the end of your nose.
Close one eye and observe where the pencil appears
to be against the background of the wall you face.
Example: if you are looking at the blackboard,
which side of the board does the pencil appear
on? How far away from the center of the board?
Now close the other eye and repeat your observations.
Be sure to state aloud what you see to your group
members. They will take note of your observations,
and then compare them to their own.

How did the pencil appear
to shift as you closed one
eye and then the other?

2. Repeat the above
process, but this time with
the pencil held at about half
an arm's length in front of
your nose. Again remember
to discuss what you see
with your group members.

At this distance, how did
the pencil appear to shift
as you closed one eye and
then the other?

3. Now go through the
process with the pencil held
at a full arm's length in front
of your nose. Discuss what
you observe with your
group members.

How did the pencil appear
to shift as you closed one
eye and then the other?
How did it change with
distance from your eyes?

4. Can you see a problem in trying to detect the parallax shift of stars that are relatively far away from the sun?

5. How does your answer to #4 help to explain why the geocentric model was believed for so long?

6. Make an analytical statement relating the size of the parallax angle to the distance an object is from the observer. (Direct, invserse, inverse square?)

7. Table 1 lists the parallaxes of a few bright stars observed by the *Hipparcos* satellite. Calculate their distances by the relation you found in #6. (The results will come out in parsecs.)

Table 1. Parallax Data from Hipparcos

Star Name	Observed Parallax (arcsecs)	Distance (pc)
Alpha Centauri	0.732	
Alpha Canis Majoris	0.379	
Alpha Aquiliae	0.194	
Alpha Canis Minoris	0.286	

8. Which star is closest to the sun? Which one the farthest?

Concept Extension

Imagine you observed stellar parallaxes from Mars. Would they differ from those observed from the earth? If so, how?

Focused Discussion: Classifying Stars by the H-R Diagram

Leader: _____   Reporter: _____

Skeptic: _____   Skeptic: _____   Skeptic: _____

Purpose: To visualize the relationship between surface temperatures and luminosities of stars.

Evolving Universe Connections: Sections 13.3 and 13.4, Figure 13.5; _Concept Clusters_: Cosmic Distances and Celestial Light and Spectra.

Skill Background: Graph using a powers-of-ten scale.

Procedure: A glance at the stars in a dark sky may strike you as overwhelming. Yet, astronomers can measure their traits well. By focusing on just two stellar characteristics—surface temperature and luminosity—we can get an inkling of how stars are different and how they are alike.

Table 1 provides a list of 25 stars. It contains both some nearby stars and some of the brightest stars in the sky. Most of the nearby stars will have unfamiliar names. Each star has its approximate visual luminosity (relative to the sun) and surface temperature (in kelvins) listed. Use the Graph Template to make the plot. Note that the horizontal axis is surface temperature, starting with the high end to the left (25,000 K) and decreasing to the right, down to 1000 K. Each small tic on the axis amounts to 1000 K. The luminosities are given on the vertical axis; note the wide range of values. Each tic mark here represents 1/10 of the interval. Plot each star at the correct combination of luminosity and temperature. Do the sun first, it's easy! Plot each star's point using its number, so you can tell which stars are where on the graph. With your completed graph, answer the following questions:

1. How are the stars arranged overall on this temperature-luminosity graph? Can you divide them into two or three large groupings?

2. The low-luminosity stars tend to be the nearby stars (otherwise, we wouldn't see them!). Where on the diagram do these stars fall? Is there any similarity about their surface temperatures?

3. Pick out any star hotter than the sun. Then choose another. Do you expect that stars hotter than the sun are generally more or less luminous than the sun?

Table 1. Properties of Selected Stars

Number	Star Name	Visual Luminosity	Surface Temperature (K)
1	Sun	1.0	5800
2	Luyten 726-8A	0.00006	2600
3	Epsilon Eridani	0.30	4600
4	Aldebaran	690	3800
5	Eta Aurigae	580	16,000
6	Rigel	89,000	12,000
7	Betelgeuse	20,000	3300
8	Mu Camelopardalis	150	3000
9	Canopus	9100	7400
10	Sirius A	23	10,000
11	Sirius B	0.003	10,000
12	BD +5° 1668	0.0015	3000
13	Procyon A	7.6	6500
14	Iota Ursae Majoris	11	7800
15	Zeta Leonis	50	8800
16	Wolf 359	0.00002	2600
17	Lalande 21185	0.0055	3300
18	Ross 128	0.00036	2800
19	Spica	1900	20,000
20	Arcturus	76	3900
21	Alpha Centauri A	1.3	5800
22	Beta Canis Minoris	240	12,000
23	Antares	3600	3000
24	Zeta Ophiuchi	4500	23,000
25	Vega	52	11,000

Graph Template

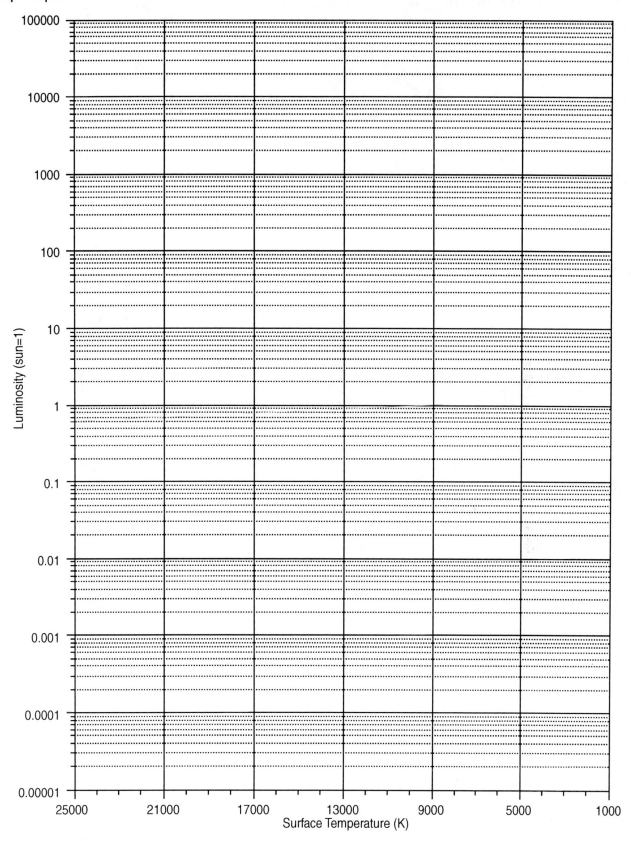

Luminosity (sun=1)

100000
10000
1000
100
10
1
0.1
0.01
0.001
0.0001
0.00001

Surface Temperature (K)

25000   21000   17000   13000   9000   5000   1000

Focused Discussion: Stellar Evolution

Leader: _____ Reporter: _____

Skeptic: _____ Skeptic: _____ Skeptic: _____

Purpose: To find an evolutionary track for people and compare it to an evolutionary track of stars.

Evolving Universe Connections: Section 15.1, Figures 15.2 and 15.3; *Concept Clusters*: Cosmic Distances and Scientific Models

Procedure: Table 1 gives the height vs. weight for a typical *group* of people in the United States. Plot the values on Graph Template 1.

Table 1. Height vs. Weight of People in the U.S.A.

Height (inches)	Weight (lbs)	Height (inches)	Weight (lbs)
60	115 lbs	68	165
62	125	70	170
64	130	71	178
65	140	72	180
66	145	74	200

Table 2 gives the height vs. weight for a *single individual* as he (a male) ages. Plot this data on Graph Template 2 and label the stages of development, such as birth, childhood, adolescence, middle age, and old age.

Table 2. Height vs. Weight of an Aging Male Individual

Height (inches)	Weight (lbs)	Height (inches)	Weight (lbs)
21	8	60	95
24	25	64	115
36	45	66	130
48	60	68	145
54	80	68	155
58	90	68	170

1. What are these two relationships that you have plotted? That is, how do they differ and what do they have in common?

2. How can you relate the two diagrams to the lives of stars and your textbook's. H-R diagram? (Hint: Time!)

Graph Template 1

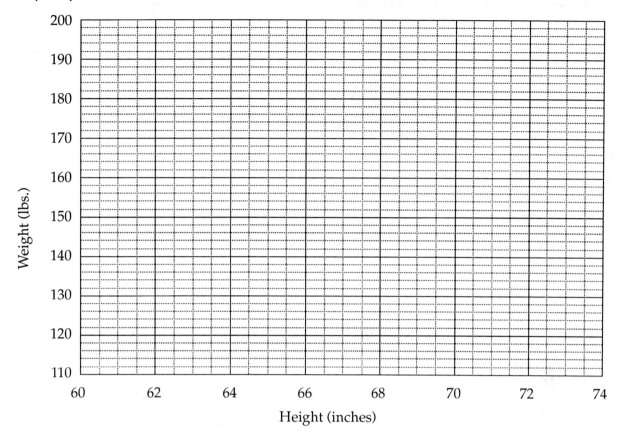

Weight (lbs.)

Height (inches)

Graph Template 2

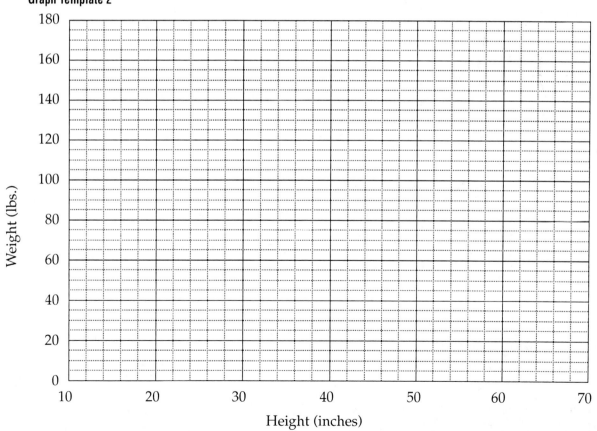

Weight (lbs.)

Height (inches)

Focused Discussion: Cepheid Variable Stars

Leader: _____ Reporter: _____

Skeptic: _____ Skeptic: _____ Skeptic: _____

Purpose: To find a relationship between the periods and luminosities of cepheid variable stars and use it to determine distances.

***Evolving Universe* Connections**: pp. 351-352, Figure 15.20, Section 17.3, Figure 17.9 and 17.10; *Cosmic Cluster*: Cosmic Distances.

Procedure: Cepheid variables are giant and super-giant stars that expand and contract. As they do so (we measure this change by the Doppler shift), their luminosities vary in a regular fashion. The time interval from peak brightness to the next peak brightness defines the *period* of light variability. The name "cepheid" comes from the star that is the pro-totype of the group, Delta Cephei (fourth brightest star in the constellation Cepheus). The cepheid vari-ability marks a stage late in a star's evolution, as it burns helium (to carbon) in its core. These stars are more massive than the sun, typically a few to ten solar masses.

Table 1 provides the periods and luminosities (to two or three significant figures) of selected cepheid variables. These luminosities are the *average* values, since these stars vary! (You will probably *not* recognize the names of any of these stars.) Use the Graph Template to plot these data. Notice that the x-axis (period in days) has an increment of 1 day; the y-axis (luminosity in solar luminosities) has an increment of 500. Once you have plotted the values for the stars, draw a "best fit" straight line through the data points. (Position the line so that it goes through most of the points, and it has about as many points above it as below it. A region of many points should influ-ence the line more than a region of few points.) Using your graph, answer the following questions:

1. What is the general trend of period versus luminosity?

2. Using your line, estimate the luminosity of a cepheid variable with a period of 25 days.

What of one with a period of 30 days?

3. Suppose you measure the flux at the earth of a cepheid whose period is 30 days. What procedure could you use to find out the distance to this cepheid?

Table 1. Period and Luminosity Data for Selected Cepheids

Star	Period (days)	Luminosity (sun = 1)
SU Cas	2.0	1000
CF Cas	4.9	1820
VY Per	5.5	2820
V367 Sct	6.3	3470
U Sgr	6.7	3890
DL Cas	8.0	3720
S Nor	9.8	4470
VX Per	10.9	5890
SZ Cas	13.6	8510
VY Car	18.9	11200
T Mon	27.0	18600
RS Pup	41.4	22400
SV Vul	45.0	30000

Graph Template

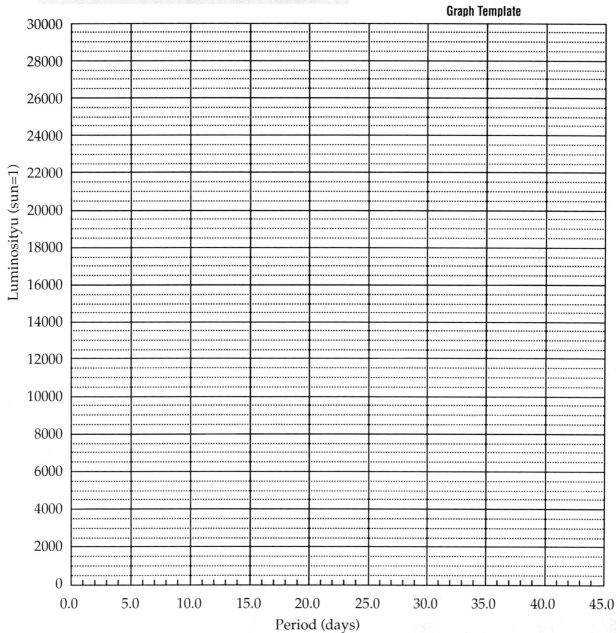

Focused Discussion: The Sun's Distance from the Galactic Center

Leader: _____   Reporter: _____

Skeptic: _____   Skeptic: _____   Skeptic: _____

Purpose: To find out how we know the sun's location in the Galaxy.

Evolving Universe **Connections:** Figures 17.6 and 17.20; *Concept Cluster*: Cosmic Distances.

Procedure: Table 1 lists selected globular clusters (given in order of name in the New General Catalog, NGC) and their three-dimensional coordinates (X, Y, and Z). X and Y are the distances in the plane of the Galaxy. Z is the distance above and below the plane. Each group will be responsible for a subset of these data. Mark "Sun" at position 0,0 on the Graph Template.

---

1. Each group plots their data (X, Y coordinates only) using the Graph Template. After the positions are plotted, draw a circle around the points and estimate the center.

---

2. Using your group's data only, calculate the distance of the sun from the center of the globulars. It is the length of the line from the sun to the globular's center.

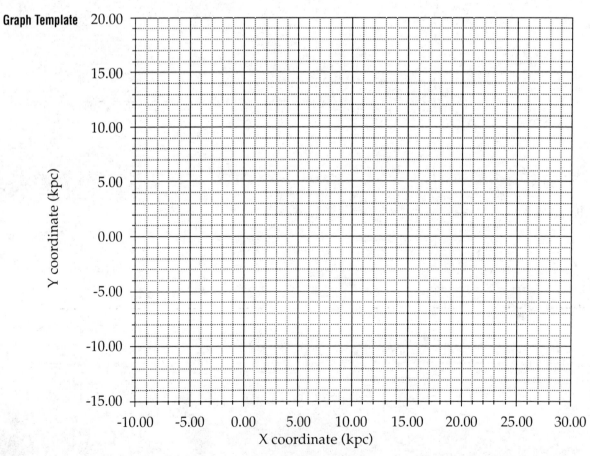

Graph Template

Y coordinate (kpc)

X coordinate (kpc)

54

© 2002 Michael Zeilik

3. Send a gofer around to the other groups to get the distance of the sun estimated from their data. Write that information here and calculate the class average in kpc.

4. Do they differ by much? If so, explain. Transform the distance from parsecs to light years by multiplying by 3.26. What is the average value for the class?

Table 1. Positions of Globular Clusters

Name (NGC) Group 1	X (kpc)	Y (kpc)	Z (kpc)	Group 4	X	Y	Z
104	3.79	-5.22	-6.44	6284	19.68	-0.54	3.48
288	-0.01	0.01	-14.04	6293	10.78	-0.44	1.51
362	5.21	-8.47	-10.39	6304	9.05	-0.67	0.86
1851	6.13	-12.85	-9.99	6316	13.71	-0.68	1.39
1904	-9.81	-9.64	-7.41	6333	8.91	0.86	1.69
2808	2.28	-10.53	-2.16	6341	3.17	8.00	6.01
3201	0.97	-7.79	1.20	6352	6.80	-2.31	-.090
4372	4.54	-7.71	-1.57	6356	16.20	1.91	2.94
4590	4.79	-8.44	7.06	6362	7.86	-5.39	-3.02
4833	4.77	-7.19	-1.21	6388	13.86	-3.56	-1.68
Group 2	X	Y	Z	Group 5	X	Y	Z
5139	5.07	-6.24	2.15	6402	7.85	3.06	2.22
5272	2.00	1.82	13.54	6440	7.52	1.02	0.50
5286	9.00	-10.15	2.54	6441	10.39	-1.30	-1.00
5466	3.16	2.85	14.48	6522	15.08	0.26	-1.03
5694	20.99	-11.33	13.70	6541	7.31	-1.38	-1.46
5824	26.00	-13.45	11.89	6624	14.23	0.69	-1.97
5897	11.39	-3.48	6.97	6626	7.15	0.98	-0.71
5904	6.53	0.45	6.96	6637	10.31	0.30	-1.88
5927	5.24	-3.46	0.54	6638	18.64	2.58	-2.38
5986	13.50	-5.74	3.47	6652	16.28	0.43	-3.28
Group 3	X	Y	Z	Group 6	X	Y	Z
6093	11.79	-1.51	4.21	6656	3.23	0.56	-0.44
6101	7.48	-6.18	-2.86	6681	18.52	0.93	-4.11
6121	3.29	-0.52	0.96	6712	9.91	4.68	-0.82
6139	13.60	-4.32	1.76	6715	21.15	2.04	-5.26
6171	9.64	0.58	4.11	6723	12.02	0.03	-3.73
6205	3.48	5.92	5.96	6752	7.89	-3.44	-4.13
6218	6.55	1.84	3.36	6779	5.45	10.55	1.73
6254	7.06	1.91	3.11	6808	6.27	0.98	-2.74
6266	10.35	-1.16	1.33	6838	3.62	5.50	-0.53
6273	7.49	-0.38	1.24	6934	13.92	17.91	-7.78

Focused Discussion: Rotation Curve of a Spiral Galaxy

Leader: _____ Reporter: _____

Skeptic: _____ Skeptic: _____ Skeptic: _____

Purpose: To visualize how the orbital speeds of a galaxy vary with distance from the center of a galaxy.

Evolving Universe Connections: Figures 18.5 and 18.16, pp. 424–425; *Concept Cluster*: Cosmic Distances.

Procedure: Recall Kepler's third law—the period squared divided by the distance cubed equals a constant value, or $P^2/a^3 =$ constant. You have used it many times. Here we again apply Kepler's third law, this time to determine the mass of a large spiral galaxy from its rotation curve. Using spectroscopic observations of the galaxy's spiral arms, the Doppler shift from rotation can be determined. Then we can find the radial velocities at points along the galaxy, from center to edge. We then construct a table of the rotational velocities at different distances from the center of the galaxy. This is called a *rotation curve*. See Table 1.

Table 1. Data for a Galaxy

Point	Radius (arcsec)	Rotational Velocity (km/s)	Point	Radius (arcsec)	Rotational Velocity (km/s)
1	-120	-170 km/s	13	2	40 km/s
2	-96	-240	14	5	80
3	-65	-290	15	10	30
4	-50	-240	16	12	80
5	-30	-150	17	14	110
6	-20	-100	18	16	120
7	-15	-50	19	21	190
8	-10	-20	20	29	170
9	-7	-40	21	66	360
10	-3	-70	22	80	390
11	-2	-15	23	96	370
12	5	10	24	107	240

You are to plot the velocity versus radius of the galaxy, thus producing a rotation curve for the galaxy. Points 1 through 12 have been plotted for you on the Graph Template provided. Plot the remaining points, 13 through 24. Draw a smooth curve through all the points.

© 2002 Michael Zeilik

1. Reach a consensus to explain the shape of the curve you have drawn. Specifically, why does it rise on one side, dip on the other, and turn over on both ends?

Calculations: All group members should be comfortable with these calculations, so make sure to discuss each step before you start punching calculator buttons.

Now choose a point located after the turn-over of the curve you plotted, and read off the velocity, V, and the distance, a, for this point.

$$V = \qquad km/s \qquad a = \qquad arcsec$$

It is necessary to convert a into astronomical units, AU, and then into kilometers. The conversion has been simplified for you: just multiply your value for a by 26 million, or 26×10^6, to get a in AU. Your new value for a:

$$a = \qquad AU$$

To convert to kilometers (km), multiply the above value for a in AU by 150 million, or 150×10^6, to get a in km.

$$a = \qquad km$$

Now you need to calculate the period of rotation for your chosen point. To do this, multiply your value for a by 6 (2π really, but close enough) and divide by your value for V.

$$P = (a \times 6) / V = \qquad seconds$$

To convert to years, divide P by the number of seconds in a year.

$$P = P / 31 \times 10^6 = \qquad years$$

For the final step, to determine the mass of the galaxy (in units of solar mass) from the two pieces of information you have painstakingly calculated you will now use Kepler's third law.

Square the value of P in years: $P^2 =$

Cube the value of a in AU (*careful*, AU, not km!): $a^3 =$

and divide P^2 by a^3 (this will yield a very small number!)

$$P^2 / a^3 = constant =$$

Finally, to obtain the mass of the galaxy in units of the solar mass, divide 1 by the constant.

$$M_{galaxy} = 1/(constant) = \qquad M_{sun}$$

Concept Summary

This activity had a lot of calculation, much more than you normally do, but what exactly was it that you *did*? Let's break it down.

1. What observations did you start with?

2. What relationship or
rule let you figure out the
galaxy's mass from these
observations?

3. What special point did
you have to pick to find
the *total* mass of the galaxy?

4. Where have you seen this
rule applied before (say locally,
 in the solar system)?

Concept Extension

If all the mass of a galaxy
were concentrated in the
center, how would the rotation curve look? Sketch it
on the Graph Template.

Graph Template

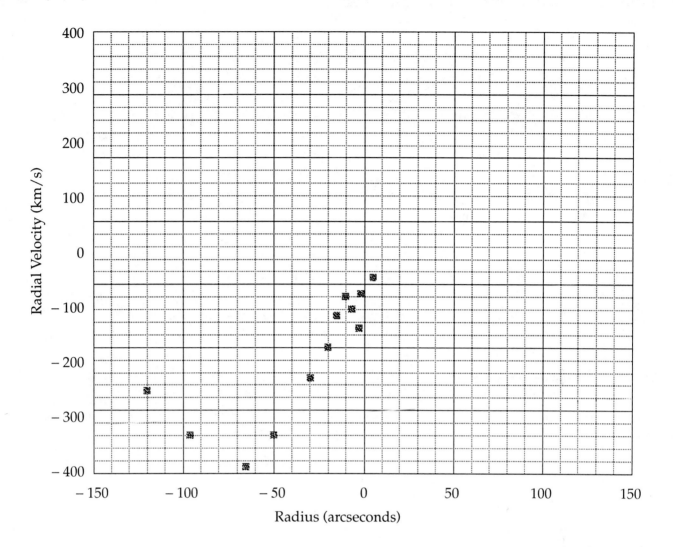

Focused Discussion: Hubble's Law

Leader: _____   Reporter: _____

Skeptic: _____   Skeptic: _____   Skeptic: _____

Purpose: To visualize the relationship between distances and recessional velocities for galaxies and find a value for the Hubble constant.

Evolving Universe **Connections**: Figures 18.13 and 18.14, pp. 419-423; *Concept Cluster*: Cosmic Distances.

Procedure: You will make "Hubble plots" of the distances and recessional velocities for selected groups of galaxies. The distances are found by a variety of ways, including the period-luminosity relationship for cepheids. The recessional speeds are found from the red shifts in the spectra of the galaxies. If this red shift is interpreted as a Doppler shift, it provides the radial velocity along the line of sight—a recessional velocity for a red shift.

1. First make a plot using Graph Template 1 from the data in Table 1. These galaxies are selected from the brightest ones in the sky (but you probably won't recognize their names). (What can you infer from the fact that they are among the brightest?) Graph Template 1 has the horizontal axis as the distance in millions of light years, from the closest to the farthest galaxies. On the vertical axis is the radial velocity in kilometers per second.

Plot the points for all the galaxies and draw a straight line through them with a ruler. DO NOT "CONNECT THE DOTS"! Try to draw a straight line so that about as many galaxies fall above and below the line as on the line. Now measure the slope of the line, which is the rise (y-axis) over the run (x-axis). Use the *complete* length of this "best fit" line, not just a part of it.

2. Find the slope from the difference in the value of the y-axis over the difference in the value of the x-axis. What value do you get? This is your value of Hubble's con-stant from the information in the graph.

3. All velocities for these galaxies are recessional—they are red shifts. What does this tell you about the universe: is it static, expanding, or contracting?

4. Now turn to Table 2, which lists a different sample of galaxies. Each one is chosen from a cluster of galaxies. Use Graph Template 2 and the plotted data to find the slope. What is it?

5. How do your values compare? Do they differ significantly? If so, why?

Table 1. Selected Bright Galaxies

Galaxy	Distance (Mly)	Radial Velocity (km/s)
Fornax A	98	1713
Messier 66	39	593
Messier 106	33	520
NGC 4449	16	250
Messier 87	72	1136
Messier 104	55	873
Messier 64	23	350
Messier 63	36	550
NGC 6744	42	663

Graph Template 1. Hubble's Law

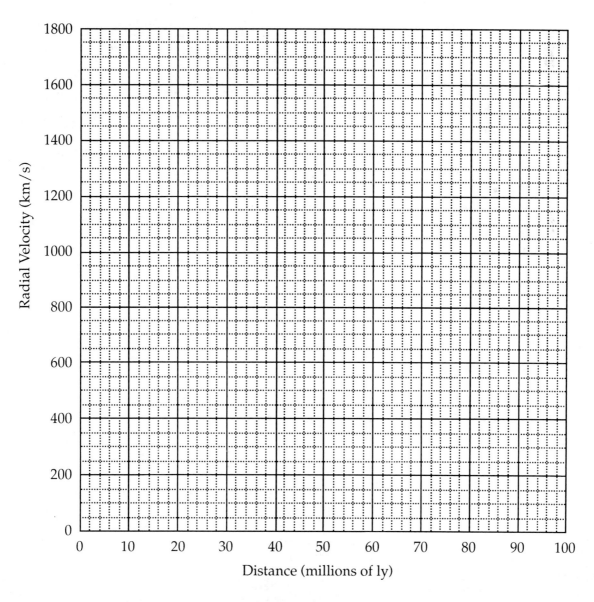

Table 2. Selected Galaxies in Clusters of Galaxies

Galaxy in	Distance (Mly)	Radial Velocity (km/s)
Virgo cluster	63	1210
Ursa Major cluster	990	15000
Corona Borealis cluster	1440	21600
Bootes cluster	2740	39300
Hydra cluster	3960	61200

Graph Template 2. Hubble's Law

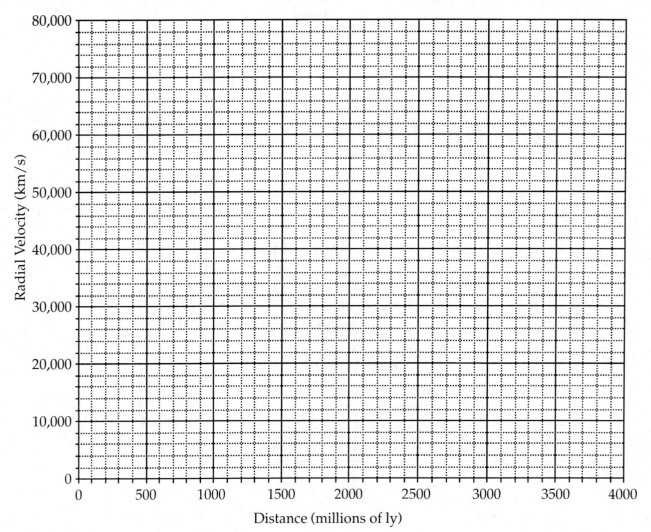

Focused Discussion: Cosmic Background Radiation

Leader: Reporter:

Skeptic: Skeptic: Skeptic:

Purpose: To be able to identify and interpret physically the spectrum of the cosmic background radiation.

Evolving Universe **Connections:** pp. 467-470; *Concept Cluster*: Celestial Light and Spectra.

Procedure: A *spectrum* consists of light spread out over wavelength. Table 1 shows a series of measurements of the cosmic background radiation. (Note that the wavelength is given in millimeters.) The values are relative to the peak, so they measure *relative intensity*. Note that the relative intensities vary over the range of wavelengths; this indicates that the source is *not* emitting the same energy per second at each wavelength. Use the data from the table to plot a curve in the Graph Template. After you have plotted all the points, draw as smooth of a line as possible through them.

Table 1. Cosmic Background Radiation Spectrum

Wavelength (mm)	Relative Intensity
0.40	0.04
0.60	0.40
0.80	0.85
1.00	1.00
1.20	0.95
1.40	0.85
1.60	0.70
1.80	0.58
2.00	0.47
2.20	0.38
2.40	0.30
2.60	0.25
2.80	0.20
3.00	0.18

1. What kind of spectrum is this?

2. At what wavelength (in millimeters) does the spectrum peak?

3. What *one* physical property does this kind of emission depend upon?

_____

Wien's law provides the relationship between the wavelength at which a Planck curve peaks and the temperature of a blackbody. It is

Temperature = 2.9 ÷ Peak Wavelength

where the temperature is in kelvins if the peak wavelength is in millimeters.

_____

4. Using your result from #2, calculate the temperature in kelvins of the cosmic background radiation.

_____

Concept Extension

If the temperature were higher, how would your graph change?

Graph Template

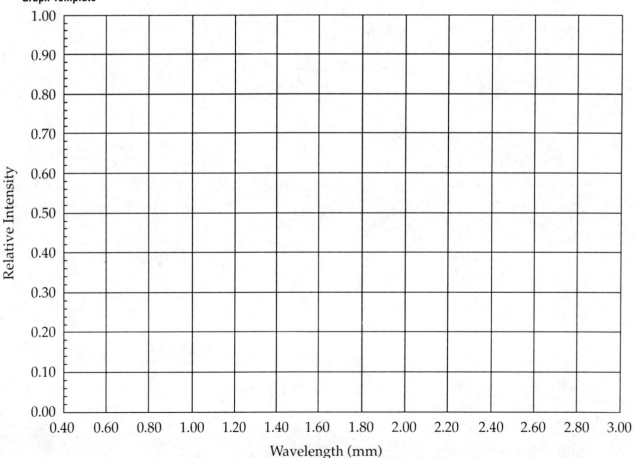

Relative Intensity

Wavelength (mm)

© 2002 Michael Zeilik